ORCHESTRATE. AUTOMATE. DOMINATE.

HOW WINNERS WORK IN THE AGE OF AI

DONOVAN GARETT

AlgoRhythms Studios, Ltd.

P.O. BOX 35643

Cleveland, Ohio 44135

United States of America

PRINT ISBN: 978-1-963267-41-9

E-BOOK ISBN: 978-1-963267-42-6

Library of Congress Control Number: 2025921410

Printed in the United States of America

Dedicated to those who make learning a lifelong ambition.

"AI won't replace humans, but humans using AI will replace humans who don't."

Ginni Rometty, former CEO of IBM

Contents

Chapter 1

Most Degrees are Dead. Let's Move On.

> *"It is not the strongest of the species that survives, nor the most intelligent, but the most adaptable to change"*
>
> Charles Darwin

FOR DECADES, A COLLEGE degree was the closest thing to a career guarantee. It opened doors, signaled competence, and gave you a head start in life. That system made sense in a world where information was expensive, gatekept, and formidable to access. But that era is over. What once made college graduates competitive now just makes them average.

AI has managed to shift the value equation of society. Being able to memorize facts or follow instructions isn't rare—it's easily replicable. Today, those skills are easily performed by software at scale. What matters now isn't how much you know, but how clearly you can think, how quickly you can adapt, and how effectively you can design successful outcomes.

College degrees don't teach that.

Systems can't measure it.

But it's precisely what the modern economy rewards.

As we begin our journey together, we'll break down why traditional credentials have lost their edge, what employers really want, and how to shift your thinking from credentialed to capable. The goal certainly isn't to tear down education. It's to expose the lie that in today's economy, having a degree (and nothing more) equates to long-term success.

A Degree Was Designed for a Different Era

Most traditional college degree programs were created to serve a very specific purpose: to prepare future employees for predictable jobs within a structured economy. At the time, having a college degree ensured that companies attracted workers who followed rules, absorbed information, and would remain loyal to a single employer for decades.

In that world, having a basic degree signaled competence, reliability, and a level of intellectual discipline. It never required innovation, adaptability, lifelong learning or leadership. You just had to prove that you could work within a system that someone else built given a defined framework.

> *"The illiterate of the 21st century will not be those who cannot read and write, but those who cannot learn, unlearn, and relearn."*
>
> Alvin Toffler

A credential-based model made perfect sense at a time when knowledge was scarce. Before the internet, before open-source tools, before rapidly evolving, on-demand training, information was real power. Large educational institutions tightly controlled the flow of it (often working directly with large corporations in exchange for sponsorships, grants, and access to top-tier jobs for their graduates).

If you wanted access to that power, you had to go through them. Now, that monopoly no longer exists. Today, the information that once cost tens of thousands of dollars and several years to acquire is available for free, in a fraction of the time. More importantly, proving your actual worth can no longer be hidden behind a piece of paper.

Even companies that used to hire strictly by pedigree are gradually shifting away from this approach. They realize that a degree doesn't guarantee competence, initiative, or creative thinking. It simply shows that you finished something; not necessarily that you learned how to think. In an economy where technology is reshaping the very nature of work, that distinction matters.

Information Is Now Cheap and Plentiful

There was a time when just having access to information gave you a significant advantage. Knowledge of certain subjects was a tightly controlled resource. That advantage is now gone. If something can be searched, summarized, or generated by AI in seconds, it no longer sets you apart. Information isn't the bottleneck anymore. Attention, clarity, and skillful implementation are.

> *If information is free, then the only thing that makes you valuable is what you do with it.*

Beginning in the mid-1990s, search engines made it possible to find any fact. AI tools now make it possible to use those facts in ways that once required professional expertise. Now, the average person can access what specialists used to provide: research, writing, analysis, design, and planning. Along with that shift, knowing things is much less valuable than knowing what to do with them.

Overall, the barrier to entry has collapsed. You don't need to sit in a classroom to learn complex skills. You don't need a degree to start creating value at scale, and employers are noticing. They care less about where you studied and more about how you think, solve problems, and communicate ideas. Career advantages no longer belong to the most credentialed. They belong to the most capable.

Employers Want Outcomes, Not Credentials

Today, companies aren't asking, *"Where did you go to school?"* nearly as often as they used to. They're asking, *"Can you solve this problem?"* Employers care about what you can produce, not what you've memorized, or what you knew at a specific point in time. Results are the new currency. And, of course, results can't be faked by a framed diploma on the wall.

Large companies who can offer high salaries are quickly shifting toward skill-based hiring. That means portfolios, test projects, simulations, and situational analysis exercises. In many cases, someone without formal education, but with a solid track record of results, is a stronger candidate than someone with multiple degrees and unproven performance. Degrees used to be a proxy for potential. That's being phased out by something more tangible: evidence.

Even in fields that traditionally required advanced education, companies are starting to bypass credentials when a candidate demonstrates clear thinking, initiative, and execution. They don't need someone who's been sitting in a classroom. They need someone who can make sense of complexity and take decisive action to achieve their stated business objectives. That shift is already in motion—and it's accelerating.

The Degree Trap: Why People Still Chase It

Despite all the recent changes in how work is evaluated, people still cling religiously to degrees. Not necessarily because they believe in their value, but because the alternative is too uncomfortable to think about. To many, a degree feels like a guarantee. It suggests safety, structure, and a strong sense of *"social approval."* For some, it's much more than just a one-time credential—it's a critical part of their identity.

From early childhood on, parents push degrees on their children because they themselves were taught that it's the only path to a stable life. The underlying problem with this thinking is that it's based on an outdated model. This isn't 1957. The so-called *"middle-class"* lifestyle that our great-grandparents enjoyed is no longer in reach. The global economy has permanently changed.

Schools continuously promote the pursuit of degrees in order to produce graduates, not independent thinkers. Employers (even ones that no longer require degrees) still include them by default in job listings. This naturally creates a kind of *"social inertia."* As a result, most people obtain degrees not because they're useful, but because they're expected to.

Of course, this is a dangerous trap.

Today's youth spend precious years and rack up thousands of dollars of debt chasing credentials that may no longer be relevant in the world they're stepping into. They delay action, avoid risk, and follow a 'structured path' in the hope that someone will eventually reward them for finishing it. But the modern economy no longer guarantees those rewards. The movement has weakened—and in many fields, it's broken entirely.

The Cost of Blind Loyalty to a Dying System

The ripple effects of chasing a degree without questioning its relevance are much more than financial. They're also strategic. To be sure, student debt will haunt you for decades. The bigger price, though, is time lost and momentum stalled. Theories such as Moore's Law suggest that the technical knowledge and specific tools learned in the first year of a degree will most likely be obsolete by graduation. So, while you're paying top-dollar to learn outdated material in a rigid, unforgiving system, others are adapting to market forces in real time.

Many graduates enter the workforce with no marketable skills, no clarity on their direction, and no real advantage in the job market. They've been trained to follow instructions, *"color inside the lines,"* and not question the system. Worse still, the structure of school rewards compliance and memorization—traits that are easily replaced by automation.

Loyalty to a system that no longer serves you isn't noble. It's incredibly risky. The assumption that degrees equal opportunity is no longer reliable. In fact, it can quietly keep you stuck—invested in an expensive framework while the world moves on without waiting for you to catch up.

Reflection: No One Asks Anymore

I distinctly remember a time when people used to ask: *"Where did you attend school?"* —like it was a part of your core identity or something. It was one of the first questions in an interview, at a networking event, or on a résumé screen. That one answer could open doors. It carried real weight.

Now?

No one asks.

Not because they're being polite or politically correct. It's because they simply don't care.

After nearly 30 years, I've learned that no one's diligently checking where you studied (unless, of course, you're planning to practice medicine, law, or a critical discipline like structural engineering).

They really want to know whether you can solve the business problem in front of them. They want clarity, not credentials. They want to know if you can think, adapt, build, and lead—because those are the critical skills that AI can't replace.

What Actually Matters Now

In today's economy, value isn't measured by stodgy pedigree. It's measured by clarity, adaptability, and execution. Employers, clients, and partners want to know if you can quickly understand what matters to their bottom line, move with intention, and produce consistent, high-quality results—especially in business environments that pivot fast and markets that change often.

The skills that matter now aren't taught in most classrooms. They include the ability to filter noise, ask better questions, work well in teams, manage your own attention, build repeatable systems, and make both tactical and strategic decisions under pressure. Technical ability helps, but it's not enough. You need mental agility—the capacity to zoom in when needed and zoom out when it counts.

A degree may help you get your foot in the door. But the ability to stay relevant now comes from an entirely different skillset.

Today, real power lies, not in what you've done at a single point in time, but in how you think and perform consistently. That means learning how to analyze inputs, use tools like AI to speed up execution, and focusing on outcomes over activities. People who do that consistently are the new standard—not the exception.

In the AI era, we will also begin to see a dramatic paradigm shift in the flow of money and wealth. We will see money transfer from the credentialed and pedigreed to the truly talented. From those who merely look good on paper, to those who create real value at scale. For many, that's a frightening concept—one that they're not ready for.

This shift won't just happen in developed countries and among the upper-class. It will cross geopolitical zones, and will possibly disrupt entire nations. Now, a 12-year-old child with a cellphone in rural India or the African bush has access to the same AI technology as a multi-national corporation.

We've hit an inflection point that we cannot turn back from.

Winners Have Already Moved On

While the status quo still clings to outdated career paths, forward-thinkers have already made the shift. They don't ask for permission to start—they build. They don't worry about formal titles—they focus on outcomes. And, most notably, they're not waiting to earn credentials—they're proving their value in real time.

You'll find them publishing content, creating products, automating workflows, and learning new tools as they go. They use AI to multiply their thinking, not just their output. Many have no formal credentials—and it doesn't matter. What matters is that they can see clearly, execute skillfully, and acclimate quickly to a constantly changing business and technical landscape.

This cohort of innovators isn't necessarily younger. They're just more aligned with current reality. They've stopped investing in systems that no longer deliver returns. They realize that in a world where anyone can learn anything, the most value-driven people are those who can apply what they learn faster than anyone else.

It’s time to stop chasing the wrong prize.

AI has changed the rules.

A degree *used to be* the key to opportunity. It gave you a head start in a system that valued credentials, routine, and compliance. That world no longer exists. Access to information isn't rare. The ability to think clearly, act decisively, and deliver value is.

What employers, markets, and clients care about now is your ability to produce outcomes, lead systems, and work with the tools that define this new era.

If you’ve been clinging to credentials as a form of identity or long-term security, it’s time to let that go. The real advantage belongs to those who can quickly adapt.

This book will help you do just that.

Chapter 2
Drowning in Data, Thirsting for Thought

"We are drowning in information, while starving for wisdom."

E.O. Wilson

FOR THE PAST THREE decades, we've been living in a world where almost any answer is one question away. Search engines, AI tools, chatbots, and content platforms have made knowledge more accessible than ever before.

But that's not the problem.

The problem is that easy access to information has created lazy minds and sloppy thought. When information is everywhere, few people stop to think and ponder. They skim, scroll, summarize—and then move on without understanding anything on a deeper level.

In this chapter, we'll talk about the paradigm shift from collecting knowledge to applying it. It's no longer impressive to know facts. AI can do that better and faster. What separates people now is the ability to process, filter, and think critically about the information they interact with.

Knowing things is common.

Understanding how to use them is rare.

You'll learn why shallow thinking is one of the biggest liabilities in the age of automation, and how to rebuild your ability to think clearly, decide confidently, and create meaningful results in a world that, unfortunately, rewards speed over depth.

Information Is Cheap. Deep Thought Is Priceless.

Whether you choose to consciously think about it or not, the sheer volume of information we're exposed to on a daily basis is staggering. News updates, social feeds, search results, emails, podcasts, ads, tutorials. The list goes on and on.

The real problem? . . .

Most of it is just noise. Very little 'information' gets turned into action.

Even less becomes actual understanding and enlightenment.

The internet removed the barrier to knowledge. Then, AI removed the barrier to using it. But neither tool guarantees clarity. In fact, for most people, the constant flood of information makes thinking harder, not easier.

Their attention is split.

Their thought is shallow.

They consume content, but they don't retain or apply it. It just passes right through them.

Being surrounded by knowledge doesn't necessarily mean you're learning anything. It means you're constantly one distraction away from forgetting what you just read. If you can't sit with an idea, question it, connect it to something else, and draw out meaning, you're not thinking.

You're just reacting.

Knowing vs. Understanding

You can memorize a hundred facts and still not understand how any of them work together. That's the difference between storing information and thinking with it. AI and deep learning algorithms can store and retrieve more data than you ever will.

True understanding allows you to make decisions with confidence. It's what helps you evaluate risks, spot contradictions, and move past surface-level answers. It's what makes you useful in meetings, credible during strategy sessions, and priceless during project execution.

Knowledge alone won't get you there.

Connection, clarity, and communication will.

The rise of tools that '*know things*' has tricked people into *feeling that they're thinking*, when, in reality, they're just *summarizing*. Sure, they can quote articles, repeat talking points, or generate text—but they don't really know what to do with any of it. That's the new gap: not between the informed and the uninformed, but between the *reactive* and the *reflective*.

If you're not challenging what you consume, you're not learning—you're hoarding content.

See, you can use AI to write answers, summarize articles, or generate solutions. That doesn't mean you understand what's behind it. Tools can synthesize. Only you can evaluate.

The danger lies in thinking that access to good information is the same as having a good mind.

It isn't.

Understanding requires friction. It takes effort. You have to really wrestle with the problem, not just copy and paste a response. You have to ask *why* something works, what it affects, and whether a given solution *should* be trusted. That's where insight comes from—not from collecting content, but from testing and applying it with intent.

AI Won't Think for You It Just Amplifies What You Think

AI can generate responses, solve problems, and simulate thinking—but only when (and how) it's directed.

AI is not a mind.

It's a mirror.

It reflects the quality of the prompts, assumptions, and intent behind it. If you feed it weak input, it gives you generic output. If you guide it with clarity, it becomes a powerful extension of your thought process.

That's where most people today get it wrong. They assume the tool will do all of the heavy lifting for them. What they fail to realize is that AI doesn't replace the need for thinking—it only multiplies the effects of whatever thought process you bring to the table. If your direction is vague, confused, or shallow, the results will be too.

This illustrates why critical thinking is more valuable now than ever. The ability to define a problem clearly, challenge your own bias, structure a prompt with purpose, and then evaluate what comes back—those are the skills that make AI powerful. Without it, you're just outsourcing your brain and hoping something of value comes back in the chat session.

Why Shallow Thinking is on the Rise

The more convenient information has become, the less effort people put into actively engaging with it. Headlines are read 500% more often than the actual article. Summaries are preferred over full chapters. Content is consumed in short bursts, optimized for speed, not depth. The result is a population that knows a little about a lot—but not nearly enough to make decisions with confidence.

> *"On the average, five times as many people read the headline as read the body copy. When you have written your headline, you have spent eighty cents out of your [advertising] dollar"*
>
> David Ogilvy

The mere fact that you're taking the time to read this book says that you're different than most. Did you know that according to a Pew Research Center study, only 54% of the U.S. population read a book in 2023? [1] That means that a whopping 46% (almost half of the country) didn't read a single book.

As you've probably guessed, algorithms feed into this behavior. They reward what grabs attention, not what builds deep understanding. Fast takes, witty captions,

---

1. https://today.yougov.com/entertainment/articles/48239-54-percent-of-americans-read-a-book-this-year

emotional reactions, and surface-level opinions dominate most online platforms. We're subtly being conditioned to skim, swipe, and scroll—not to sit, question, or reflect. The mental muscles that once supported deep thinking have atrophied.

As a society, we seem to be getting dumber.

This isn't just a personal fallacy—it's a professional liability. In work environments where clarity and strategy are essential, shallow thinkers are quickly exposed. They commonly repeat opinions, overreact to noise, and struggle to explain their logic and reasoning. When everyone has access to the same tools and data, your edge depends on how well you can think.

Critical Thinking and Focus Are Your New Superpowers

In an economy where tools are easily available and information is free, the only rare skill left is clear thinking. It's the ability to pause, take a step back, ask the right questions, cut through the chatter, and reach useful conclusions. That skill can't be outsourced, automated, or copied without effort. It must be developed over time with practice.

Critical thinking doesn't mean being skeptical of everything. It's knowing how to structure your thoughts, assess assumptions, and comfortably navigate uncertainty. It's the discipline of thinking before reacting, choosing questions over 'knee-jerk' answers, and challenging the first idea that may come to mind.

> *If your conversation responses are lightning fast, or clearly driven by emotion, you're not thinking—you're reacting. In a world full of reactive people, intentional thinkers stand out.*

AI can generate a thousand options in milliseconds. Critical thinking tells you which one makes sense. Data can quickly overwhelm you. Critical thinking tells you what matters and why. When distractions are constant and opinions are cheap, those who stay focused, filter effectively, and act deliberately become the clear winners.

Train Your Brain to Think Again

Most people assume deep thinking is automatic.

It isn't.

It's a skill.

And, like any skill, it weakens when neglected.

Naturally, when it comes to thinking, shallow input leads to shallow output. If you want to think better, you have to recondition your mind to tolerate depth, discomfort, and complexity again.

First, slow down.

Avoid jumping from task to task, article to article, app to app.

Give your brain sufficient time to work through one idea or one problem before you latch on to the next one. Read without multitasking. Take notes by hand. Pause when something confuses you, rather than glossing over it.

Next, take a moment.

Write out your thoughts. Organize them in a journal or notebook. Challenge assumptions. When you hear something you agree with, ask yourself: *"Why?"* When you hear something you don't agree with, ask: *"Is my reaction emotional or well-reasoned?"* Critical thinking doesn't happen by accident. It happens when you deliberately create space for it and build mental muscle over time.

Reflection: Who's Thinking—You or the Headlines?

Look at how much of your input comes from news, social media, and headlines designed to provoke, not inform.

- Are you responding to what's *true* or what's *loud*?
- Do you find yourself repeating political or social talking points without fully understanding the context?
- When you consume news, do you carefully analyze what's being said—or who benefits from you believing it?

Modern media doesn't reward nuance. It rewards outrage, urgency, and division. If you want to think clearly, you have to protect your mind from systems built for shallow thinkers.

Winners Don't Just Consume—They Distill

These days, anyone can gather information. But very few know what to do with it. The people who move ahead aren't the ones consuming the most. They're the ones who know how to distill complexity into clarity. They don't just absorb knowledge; they transform it into something valuable.

Winners ask, *"What's the takeaway here?"* every time they read, listen, or watch. They look for patterns. They extract the three main points that matter and then discard the rest. While others drown in detail, they draw insightful conclusions.

This skill, distillation, helps people lead with clarity. It's how strategy is formed, how communication stays clear, and how action remains focused. In a world full of overwhelmed minds, the ability to simplify without losing context is one of the sharpest advantages you can develop.

Access to knowledge used to be a competitive edge. Today, everyone has it at scale. What truly separates people is not what they know—but how they think. Those who rise are the ones who slow down, filter out biases and assumptions, and make decisions rooted in clarity.

Thinking is no longer optional.

In a world where everyone has the same tools, it's how you direct them that really counts.

Chapter 3

Silence Noise, Sharpen Your Mind

"Your focus determines your reality."
George Lucas

HAVING ACCESS TO THE latest tools and information means nothing if your attention is shattered. Most people aren't failing in life because they lack intelligence. They're really failing because they can't focus. They live in a state of constant interruption. They're constantly pulled from one app to another, one message to another, one screen to another—until their mind becomes too scattered to think clearly.

In this chapter, you'll learn how to take back control of your mind. We'll discuss how to create space for real work, deep thought, and meaningful progress in a world designed to keep you distracted. 'Noise' isn't limited to what's physically around you—it's much broader—it's what fills your head and keeps you stuck in a cycle of shallow action.

You'll see why distraction has become a default societal behavior, how overstimulation disguises itself as productivity, and what practical steps you can take to clear out the mental clutter.

Today, your best advantage isn't working harder. It's sharper thinking.

You're Not Lazy. You're Overstimulated

Most people assume they need more 'motivation.' They think that their problem is laziness. It's not. It's overstimulation. Your brain isn't failing you. It's reacting exactly as it's been trained. Constant pings, notifications, headlines, comments, videos, and dopamine spikes have conditioned your mind to seek novelty rather than depth.

Our modern, tech-driven world is designed to keep our attention scattered. As soon as something feels hard, something easier shows up to replace it. Swipe. Scroll. Click. Over time, your brain adapts to this pattern, and craves the relative ease required to complete (what feels like) 'work.' Focus starts to feel like friction. Deep work becomes uncomfortable. Quiet feels awkward.

This isn't a personal failure; it's a predictable outcome of modern life. However, if you want to deliver real value, or lead with clarity, you can't work in a state of mental fragmentation.

Ultimately, the solution isn't more motivation. It's less distraction.

Distraction by Design

We often don't realize that our environments are not neutral. They're meticulously designed to interrupt us constantly. Every app on your phone, every tab in your browser, every platform you check has one goal: hold your attention long enough to benefit someone else. In fact, you don't have to look for things to distract you. They appear constantly throughout the day.

> *Distraction is not a glitch in the system—it is the system.*

The more you give in and respond to interruptions, the more your brain learns to expect them. You start working in fragments. You start to lose track of what matters. Even when you have time to focus, your mind remains in a scattered state.

This pattern can't be overcome with sheer willpower. It must be confronted head-on. Tackling the problem requires design and intentionality. Until you take control of what enters your mind, you'll stay stuck in a cycle of reacting to whatever shows up next. Focus isn't simply a habit—it's a series of conscious decisions that must be managed day-to-day.

The High Cost of a Scattered Mind

When your attention is fragmented, everything suffers—your work, your decisions, your ability to follow through. You might stay busy, but you stop making meaningful progress. You start projects that you never finish. You overthink small problems and overlook big ones. You begin to confuse motion with momentum.

Scattered focus leads to shallow work. You can't solve complex problems when your thoughts are constantly interrupted. You can't lead with confidence when your mind is cluttered. And you can't grow if every new notification pulls you in a different direction before you've seen the last task through.

The real cost of distraction isn't just *time* lost—it's *thinking power* lost. When your mind is scattered, even simple tasks take longer. Your tolerance for mental effort drops.

You lose the ability to really sit down with with your ideas long enough to do anything meaningful with them. That's not just inefficient—it's unsustainable.

Winners Build Bandwidth

Have you ever watched a successful person carefully? They just seem to move differently through life, don't they?

This isn't a coincidence.

While most people try to get more done by working faster, winners take a different approach. They build space around their thinking. They don't just manage their time. They also manage their attention. They eliminate what doesn't matter so they can stay focused on what does.

Our mental bandwidth is limited. Every distraction, every decision, everything that we think about, focus on, or try to mentally absorb takes a slice of it. So, if your day is filled with clutter (digital or otherwise) there's no room left for strategy, focus, or productive output. The sharpest minds don't operate under pressure all the time. They protect their mental capacity because it's their greatest asset.

In the age of AI, productivity doesn't mean squeezing more into your day. It means eliminating noise, using leverage, and creating adequate mental space for higher-level thinking.

You can't make good decisions if your head is full of clutter. You can't lead with clarity if you're chasing ten things at once. Focus isn't simply doing deep work. It also means fiercely protecting the environment that makes deep work possible.

The Mythos of Being Always Available

Being available all the time to everyone is commonly mistaken for being a responsible adult.

It's not.

Constant availability, or, always *'being on,'* fragments your attention, breaks your momentum, and subtly trains everyone around you that it's perfectly acceptable to interrupt your workflow without any consequences. It essentially turns your day into a steady stream of small asks and shallow responses, leaving no margin for deep work or mental clarity.

It's important to note that *availability* isn't the same as *reliability*. The people who get the most done aren't the ones who respond instantly to every single request. They're the ones who protect their time, guard their mental space and then deliver consistent, high-value results. They set realistic expectations, manage communication on their terms, and make themselves unavailable for all the right reasons.

In a world where urgency is often manufactured, productive people operate with razor-sharp intention. Their minds don't constantly check, reply, or react to every message or *'pop-up'* notification. They build systems that allow them to stay focused on what matters most, and they teach others to respect those boundaries.

It's Time to Repossess Your Mind

Taking control of your attention isn't complicated, but it does require self-discipline. Start by turning off any notifications you don't need. Not muting. Not silencing. Turning them off. If something really matters, you'll find out without being *'pinged'* every few minutes.

Block some time to think (not just to work). Give yourself a little space where nothing is pulling at you. No tabs open. No phone nearby. No app notifications flooding in. Just space to work through ideas, structure your thoughts, and direct your energy with purpose.

Replace passive habits with active ones. Rather than scrolling to decompress, read something that challenges you. Instead of clicking without any real thought, write down the challenge you're trying to solve. You don't need more "productivity hacks." You need fewer distractions and stronger boundaries.

Reflection: Winners Protect Their Mental Space

Look closely at people who consistently perform at a high level. They're not the ones who say 'yes' to everything. They're not always reachable. And they definitely don't try to please everyone around them.

They meticulously guard their schedule. They control their environment. They treat their attention like it's finite—because it is.

You can't think strategically, lead effectively, or build anything meaningful if your brain is being pulled in ten different directions.

Distraction is expensive.

Focus is leverage.

Noise Isn't Just Digital—It's People, Too

Distraction doesn't always come from devices. It also comes from the people around you (especially the people you love). Constant interruptions, emotional drama, low-value conversations, or unclear expectations. These all create *'mental drag.'* They constantly eat away at your focus and emotional energy the same way a screen does.

Winners pay close attention to their environment. They carefully monitor who adds clarity and who brings chaos. They're not rude, but they're highly selective. They don't allow unfiltered access to their time, and they don't tolerate bedlam disguised as urgency.

Protecting your mental space means screening anyone that is allowed into your mental space. This includes both their physical presence and their conversations. If you're surrounded by people who pull you off course, then you'll spend more energy recalibrating than creating. Focus isn't simply what you block out. It's also who you let in.

Focus Is a Superpower. Guard It.

Make no mistake. This world is designed to distract you. So staying focused isn't just a 'nice-to-have' habit—it's a critical advantage. While others react to every 'ping,' protecting your mental space means that you have the opportunity to step back, think clearly, and control who gains control of your attention, and when.

Remember, the real threat isn't laziness. It's overstimulation. It's the myth that you have to be always available to everyone. It's the lie that more apps and better 'productivity hacks' will make you productive. They won't—unless you first create the mental space to utilize them with intention.

Distraction is the enemy of clarity. In the age of AI, clarity is everything.

Protect it at all costs.

Chapter 4
The Conductor vs. The Soloist

"If you want to go fast, go alone. If you want to go far, go together."

African Proverb

MOST OF US WERE trained to be soloists. You know the drill: Do your part. Stay in your lane. Master your task. Rinse. Repeat.

Of course, this cycle worked well when the ultimate goal was predictability. At a time when doing your job well meant repeating the same steps with consistent results, which improved slightly over the years.

But that model is being systematically dismantled. But the world is no longer predictable. The global economy, business climates, financial markets—are all rapidly changing in real time. On top of market volatility, AI and automation are faster, cheaper, and better at task work than humans. But what they can't do effectively is lead the performance.

In this chapter, we get *'down to brass tacks.'* Here's where we begin the shift from being the one who *does* the work to the one who *directs* it. The conductor doesn't play every instrument. He coordinates the outcome. That's your new role if you want to stay relevant. The future won't continue to reward people who merely 'execute steps.' It will reward those who design the system, align the tools, and produce tangible results using leverage.

You'll learn why the soloist mindset leads to burnout and eventual obsolescence, how to start thinking like a conductor, and what it takes to shift from isolated execution to strategic orchestration. Your value isn't in how much you can do—it's in how well you direct what gets done.

The Old Way: 'Soloist' Thinking

For decades, the workplace generously rewarded people who could specialize within a given domain, stay focused, and work independently. Being a top performer in that environment meant being the best at a specific skill—whether it was writing code, processing invoices, designing layouts, or managing logistics. You were hired to do a single thing, and do it well.

This mindset, 'soloist' thinking, subtly reinforced the idea that real value comes from mastering tasks. You were expected to carry your weight, hit your numbers, and not interfere with others. Even a leadership role usually meant doing more of the same thing, just with more pressure.

That old model is obsolete. AI can perform repeatable tasks faster and at scale. It doesn't burn out. It doesn't need breaks. It doesn't ask for a raise. So if your value is centered around being a skilled soloist, you're now competing with something that can outperform you at scale without effort.

Your primary focus needs to change from the goal of working harder or protecting your task using 'tribal knowledge.' The goal is to evolve beyond it.

The New Way: 'Conductor' Thinking

If you've ever attended an orchestra, you'll notice something interesting: the conductor's name is on the marquee. Not the performers. The headlines state: "Conducted by . . ."

But here's the rub . . .

The conductor doesn't touch a single instrument.

They don't play a single note.

What they do is shape the result. They align the timing, coordinate the effort, direct the orchestra's focus, and ensure that everything works together seamlessly to produce something beautiful. Something that transcends simply being cohesive. A piece of music that is a true work of art.

That's the same mindset winners adopt in today's AI-driven economy.

Thinking like a conductor means mentally shifting from execution to orchestration. It means understanding how all the pieces connect, even if you're not the one handling each step. It requires perspective, not just raw skill. Your value comes from your ability to see the full picture and guide others (or automated tools) toward a better outcome.

In this model, you're not trying to be the smartest person in the room. You're not even the one building 'the thing.' You're trying to be the one who creates clarity. You decide what needs to happen, in what order, and with what tools. You step back from the mechanics so you can lead the performance.

Why 'Soloists' Are Quickly Being Replaced

AI doesn't need to understand the big picture to outperform a soloist. It just needs a repeatable task. That's why soloist roles are the first to be automated. Data entry, formatting, copywriting, scheduling, even some forms of analysis—these are now being handled by systems that never get tired and never forget.

Don't believe me?

Earlier this year (in 2025), Salesforce CEO Marc Benioff stated that the company reduced its customer support workforce by approximately 4,000 jobs, shifting from 9,000 to about 5,000 support employees due to the deployment of AI agents for customer service and other functions. This reduction in headcount would allow Salesforce to handle tasks more efficiently and reduce support costs, with a new "Agentforce" of customer service bots working alongside human supervisors.

> *"If we were having this conversation a year ago and you were calling Salesforce, there would be 9,000 people that you would be interacting with globally on our service cloud, and they would be managing, creating, reading, updat-*

ing, deleting data," he said, adding that those same interactions are still happening today, but "50% are with agents, 50% are with humans."

Marc Benioff, Salesforce CEO

These AI agents now work in hand-in-hand with human supervisors to manage customer interactions. The company described the move as a "re-balancing" of its workforce, where the *"AI-driven efficiencies led to fewer human support roles."* [1]

These types of headlines aren't a real concern if you're willing to evolve. But they are a dire warning. If your role revolves around executing instructions, you're in a vulnerable position. The more clearly your job can be outlined step-by-step, the more easily it can be replaced by automation or outsourced to software.

Conductors don't have this problem. They aren't necessarily measured by how many things they physically touch. They're measured by the quality and consistency of the results they lead. Or, their influence in achieving those results. Their value lies in how they utilize tools,

---

1. https://www.foxbusiness.com/economy/salesforce-cuts-4000-jobs-due-ai-ceo-says

people, and systems—not in how much they personally produce.

Trying to outwork AI is a losing game.

Learning how to direct it is how you win.

Reflection: I Watched the Typewriter Die

I'm probably dating myself here, but I was just a kid in the early '80s when the personal computer started showing up in offices. Back then, you could hear the familiar *'clickety-clack'* of typewriters everywhere you went: banks, government offices, hospitals. The cacophony of phones ringing, typewriters, and the bustling chatter of secretaries answering phone calls was everywhere.

Back in those days, secretaries were trained strictly on typewriters. They were fast, efficient, and proud of it. Then the first PCs rolled in. It didn't seem like much of a threat at first—just a new machine with a green blinking cursor.

Fast forward a year or two, and things started changing fast. The secretaries who refused to learn how to use WordPerfect or Lotus 1-2-3 were replaced quickly, quietly, and in waves. They weren't bad at their jobs. They just didn't adapt. They were still doing the same thing they'd always done, but the value of their skills had collapsed almost overnight.

I think about that now, watching people cling to every letter of their old job descriptions while AI reshapes the world around them. The pattern is the same. The tools are different. The pace sped up. The people who survive are the ones who stop doing what's comfortable and embracing what comes next.

What a 'Conductor' Actually Does

A conductor doesn't get paid for *'doing the work.'* But they are intimately knowledgeable about how the work should be done, when it should happen, who (or what) should do it, and why it matters. They think in terms of outcomes, not steps. They don't just ask what needs to get done—they ask what's worth doing in the first place.

Conductors also don't micromanage. They simply design the workflow. They decide what gets automated, what gets delegated, and what requires their direct oversight. They make sure nothing important falls through the cracks, and they keep the whole system aligned toward a result.

This type of mental shift isn't simply about leadership. It's more about leverage. Conductors create clarity out of chaos. They convert scattered efforts into coordinated movement. In a world where tools can execute with speed, your real power lies in directing those tools toward a result that matters—and ensuring all the pieces perform cohesively to achieve it.

Transformation Is Strategic, Not Tactical

Becoming a conductor doesn't require using the newest, flashiest buzzwords, tools and platforms. It means fundamentally thinking differently. It requires taking a step back from the comfort of *doing* and learning how to *lead* with design, judgment, and intent. That's a harder shift than most people realize—not because it's *complex*, but because it's *uncomfortable*.

From childhood, most of us were trained to equate effort with value. If we're not actually doing the work, we tend to feel guilty or disconnected from it. But the needed mindset shift is internal. It's understanding that doing everything ourselves is a limitation, not a strength. The real mental hurdle is being comfortable about letting go of control without losing accountability.

Remember, you're not surrendering your accountability. You're just taking ownership at a higher level. You stop measuring your worth by the number of things you touch and start measuring it by the quality of outcomes you guide. That kind of change doesn't start with a new app. It starts with how you see your role within the big picture.

Leverage Is the New Labor

In past decades, your value (and, to some degree, your potential) was directly tied to how much you could do by yourself. How many hours you could work, how many tasks you could complete, how much you could carry on your own. This paradigm rewards effort over strategy. But in the age of automation, it all falls apart. Speed and scale now come from leverage—not labor.

Leverage means using tools, systems, and people to multiply your impact without multiplying your workload. It's when one hour of your time creates ten hours of output. Or, when one decision you make improves a process that runs every day for years. Or, when one system you build replaces a hundred repetitive tasks.

This is what conductors do. They use leverage by default. They don't ask, *"What can I get done today?"* They ask, *"What can I design that keeps getting done tomorrow without me?"*

That one question changes everything.

It moves you from mere worker to strategist—from task handler to outcome driver.

How to Start Thinking Like a Conductor

Shifting into a conductor role doesn't require a promotion, or a fancy title from someone with an even fancier title. It requires a new perspective on your work. Start by focusing on the outcome. Not the steps. Not the task. Ask: *"What am I responsible for producing, and what's the smartest way to get there?"* That question alone immediately changes your posture from that of a performer to a planner.

Next, look for the strategic handoff points. What parts of this process could be automated, templatized, or delegated? You don't have to do everything yourself. You need to design a system that gets the work done reliably—even when you're not the one doing it.

Finally, zoom out. Review your week as if you were managing a team, even if that team is just you and a few tools. Ask: *"Where did my time go?" "What could be removed?" "What didn't need my direct involvement?"* Conductors operate at a higher level of control, clarity, and coordination. That perspective can be learned—and once it's mastered, you never want to go back.

Don't Just Do the Work. Direct It.

The workplace, as we know it, was built for soloists. But the workplace of the future belongs to conductors. If your entire value depends on completing task-based work, you're competing with automation. If your value comes from direction, design, and decision-making, you're in control of your role, even as the tools evolve and get smarter.

Soloist thinking is no longer enough to get by.

Leverage replaces labor.

Orchestration replaces effort.

And the highest value comes from guiding the work, not grinding through it.

You don't need to do everything yourself to stay valuable. You need to be the one who makes sure everything is moving in the right direction.

That's what conductors do.

That's what the new economy rewards.

Chapter 5
Orchestrate — Design Your Workflow

"Give me six hours to chop down a tree and I will spend the first four sharpening the axe."
Abraham Lincoln

MOST PEOPLE JUMP STRAIGHT into their day by asking, *"What do I need to do?"* Conductors start by asking a different question: *"What outcome do I need to create, and what's the best way to get there?"* This contrast is subtle—but it separates people who stay stuck in task mode from those who build systems that work at scale.

In this chapter, we'll talk about how to orchestrate your work before doing it. This is the point where you stop reacting and start designing. You'll learn how to map your outcomes, align the right tools and people, and structure your workflow to reduce chaos and create consistency. The goal here isn't necessarily to get more done. It's to build a repeatable process that delivers results without burning you out.

Throughout this chapter, you'll see why most people waste precious time and emotional energy. Not because they're disorganized, but because they never took time to properly design the work. Orchestration is how you take control of time, tools, and output before any of it slips away.

Quit Starting with Tasks

Most people start their day with a cup of coffee and a list of 'to-do's': Send these 12 emails. Make these 7 calls. Review these 5 documents. The list keeps them *busy*, but not necessarily *productive*. The underlying problem with working strictly from a list is that it creates the illusion of progress while pulling focus away from the bigger picture. As a result, we tend to think of productivity as a game of *'checking boxes,'* regardless of whether those boxes lead to a constructive outcome.

When your day begins by focusing on tasks, you lock yourself into *'execution mode'* before you've clarified direction. You move quickly without asking if you're going in the right direction. Over time, this generally leads to inefficiency, burnout, and a constant feeling of being behind—even though you're technically *"getting things done."*

Orchestration means that you flip that pattern. You start off by asking: *"What result needs to happen?"* Then you identify the smartest, most efficient way to reach that result (with or without your direct involvement). This changes how you prioritize, who you involve, and which tools you rely on. With this approach, 'doing more' is not the focus. It's designing work that gets done right.

First, Map the Outcome

Before you commit to any single task, you need to know what success looks like. Not in vague terms, but clearly, specifically, and measurably. Ask: *"What is the outcome I'm trying to produce?" "What does 'done' actually look like?"* Until you can define the ultimate objective, everything you do is guesswork.

Once the outcome is clear, 'reverse-engineer' it. Ask yourself: *"What are the major checkpoints between where I am, and where I need to be?" "What are the dependencies?"* And, more importantly: *"What steps can be automated, delegated, or eliminated?"* Mapping the entire workflow upfront saves rework, confusion, and scope creep later on in the process.

This is how conductors operate. They don't guess or bluff their way forward. They chart out the final result before making a single move. In business, most wasted time doesn't come from bad execution. It comes from unclear direction. When you map out the outcome first, the work becomes sharper, faster, and easier to replicate.

A Solid Workflow Is Your Competitive Edge

Anyone can complete a task. Very few can systematically create a process that runs smoothly and completes itself. That's the difference between *effort* and *planning*. A good workflow eliminates guesswork, reduces mistakes, and produces consistent results, even when the person executing it changes.

When your work follows a structured pattern, it's easier to scale, delegate, and improve. You stop wasting energy trying to remember what to do next or how to fix preventable issues. You simply follow a design that is already proven. If it breaks down, you know exactly where to look.

This is what distinguishes professionals from amateurs, who are constantly overwhelmed. It's not a lack of talent. It's a lack of structure. A solid workflow provides control, flexibility, and breathing room to lead instead of chase. The sharper your system, the less you'll depend on willpower or memory—and the more value you can produce under pressure.

Tools Don't Create Clarity—You Do

Too many people start with tools. They search endlessly for the perfect AI prompt, the newest app, platform, or software suite, desperately hoping it will fix their disorganization or speed up their work.

The problem? . . . Tools don't create clarity.

They only reflect the clarity you already have. A confused workflow buried inside a sophisticated system is still, at its core, a confused workflow.

You can't delegate good design to software. The real structure has to come from you. This means understanding the answers to questions like: *"What needs to happen?" "In what order?" "With what level of quality?"* Before automating anything, you must answer these questions. Otherwise, you're just moving faster in the wrong direction.

It's tempting to think that switching or upgrading tools will solve productivity problems.

Sorry, but it won't.

Until you fully understand the logic of your workflow, no tool will save you from friction. Why? Because clarity comes from thinking before building. Not from clicking before planning.

Reflection: I've Seen Crappy Software Sold to Very Smart People

I've watched million-dollar organizations buy horrible software. Not because they were dumb, but because they didn't have a fundamental understanding of their own workflow. I've seen teams buy expensive tools that created more problems than they solved—all because someone in a suit gave a polished slide deck and a controlled demo at their facility.

Now, of course, the problem *wasn't* the software. It was the fact that the people buying it didn't know what they actually needed. They couldn't describe their own process. So, they let a salesperson define it for them. Once they signed the contract, their entire workflow had to be reshaped to fit the tool's limitations.

Unfortunately, that's what happens when you don't start with clarity.

You end up changing your workflow to fit someone else's product, or vision.

Inputs, Outputs, and Feedback Loops

Every system has three core components: inputs, outputs, and feedback. Most people focus solely on the middle—what they're doing. But real control comes from understanding what goes in, what should come out, and how you measure whether the process is working.

Inputs are everything you need before work begins: information, context, tools, and expectations. If your inputs are sloppy, your workflow starts at a disadvantage. Outputs are the deliverables. Not just *"done,"* but *"done to a defined standard."* If you can't define what that looks like, don't expect consistency or quality.

Feedback is what closes the loop. It tells you whether your system is actually working. If something breaks, don't just patch the task. Instead, address the underlying issue. Trace it back. Find out what happened. Ask yourself:

"Was the input unclear?" "Was the output not defined?" "Did you skip the review process?" Focus on fixing the system, not just the symptom.

Reduce Self-Reliance

If your entire workflow falls apart when you're not there, you don't have a system. You have a critical dependency.

That's not leverage.

It's a liability.

The more your process depends on your memory, presence, or direct involvement, the more fragile it is.

In an orchestra, the members can play collectively without the conductor. Just not as smoothly (and, to wit, with no accountability for the end result).

Likewise, in business, conductors don't make themselves the bottleneck. They build repeatable workflows that can be handed off without confusion. Whether it's creating templates, setting clear standards, and documenting steps that others (or even AI tools) can follow. You don't need to be irreplaceable in every step, but you do need to make the outcome consistent and with high-quality.

Designing a process that works without you doesn't mean you're divorced from the entire process. It's more about leveling up. When you reduce the need for constant oversight, you create space to focus on the decisions and direction that actually move things forward.

Visualize the Work Before It Even Starts

If you can't see your entire workflow, you can't manage it. And, unfortunately, most people skip this step. They dive headfirst into 'the work' before understanding how everything connects. It's the business equivalent of: *'fire,' 'ready,' 'aim.'* As a result, they miss dependencies, duplicate efforts, or build in the wrong order. Visualizing the work up front gives you the perspective to spot problems before they happen.

You don't need complex software to do this. A whiteboard, a flowchart, a spiral notebook, mind map, or a simple outline—all can help you lay out what needs to happen and how it moves from step to step. What's upstream? What's downstream? What breaks if something changes? Clarity here saves time later.

When you design before you execute, you make smarter decisions about resources, tools, and timing. You avoid chaos. You can detect friction early. You move with confidence, instead of guesswork. That's what orchestration looks like in practice: seeing the whole sheet of music before playing a single note.

Design the Flow Before You Hit 'Go'

Most people waste precious time, energy, and resources because they start moving without a clear path. They jump into tasks before defining outcomes, skip structure for speed, and rely on memory and 'gut feeling' instead of clearly articulated design. That approach might feel productive at first—but it's built on a foundation of chaos.

It's time to break that pattern. Begin with the outcome, reverse-engineer the steps, and structure your inputs, outputs, and feedback loops. No AI tool can replace mental clarity. Visualizing your work prevents breakdowns, and reducing your own involvement creates more freedom, not less responsibility.

Orchestration means thinking before doing. When you build the flow first, execution is faster, cleaner, and easier to scale.

Chapter 6
Automate — Multiply Output, Not Effort

"The bad news is time flies. The good news is you're the pilot"

Michael Altshuler

YOU CAN'T INFINITELY SCALE by simply 'doing more.' We all have a finite amount of time, energy, and attention. Eventually, each of us will hit a limit. A wall, if you will. That's where many professionals stall. Sure, they're talented, but stretched dangerously thin. Some are on the brink of a complete nervous breakdown, cleverly disguised under the veil of being a *"hustler"* or *"boss chick."*

The solution to this isn't squeezing in more work. It's using smarter systems.

The focus of automation isn't taking shortcuts. It's building processes that free your mind to focus on what truly matters.

In this chapter, we'll break down how to use automation to remove friction, reduce repetition, and eliminate the need for constant manual supervision. You'll learn what to automate, when to keep things manual, and how to avoid the subtle trap of overengineering your workflow.

If orchestration is about designing the system, automation is about removing yourself from the parts that don't require your direct focus to operate smoothly. The goal isn't to escape the work—it's to focus on the parts that best utilize your unique talents, skills, and abilities.

Why Automation Isn't Optional Anymore

If you're still completing all of your tasks by hand, you're definitely not innovating – you're really falling behind the curve. Not because you're doing it wrong, *per se*, but because the entire game has changed. Tools can handle scheduling, reminders, follow-ups, data entry, reporting, content generation, customer communication, and more. Totally ignoring them doesn't make you principled. It just makes your workflow outdated.

Automation is more than just speed. It brings with it consistency, accuracy, and the ability to shift your attention to higher-value work. In competitive markets that change frequently, that's the real edge. While others burn time on tasks that should have been systemized, you're solving problems, building strategy, and moving faster without working harder.

This is no longer a futuristic concept. Think of it as your first move. The people who win aren't the ones who slog through every task. They're the ones who have figured out how to remove operations that don't need their constant attention.

Protect Your Headspace

Every decision you make and every manual task you complete drains mental energy. And it's not just time slipping away—it's also your focus. When your day is filled with repetitive actions and low-value decisions, you leave very little mental capacity for the work that actually requires deep thought.

You may recall that Apple co-founder Steve Jobs famously wore the same black mock turtleneck and jeans every day. Why? Because this simple form of 'automation' eliminated the mental effort necessary for choosing an outfit daily. In turn, this saved his decision-making 'battery' for more critical issues.

"Innovation is saying no to 1,000 things."
Steve Jobs

Automation protects your limited attention. It removes the mental clutter caused by doing the same thing over and over again. It gives your mind space to think clearly, solve real problems, and approach complex challenges with a full tank—rather than scraping by on 'mental leftovers.'

Now, this isn't avoiding hard work. It's preserving your ability and focus to do high-value work. When your calendar books itself, your status updates send automatically, and your follow-ups run on schedule, you're not just efficient—you're creating the mental space needed to control your day with clarity.

Reflection: Why I Still Use Pen and Paper

I've worked in high-pressure environments for decades—fast deadlines, big decisions, constant input. And I still carry a spiral notebook or hardcover journal everywhere I go. Not because I'm stuck in the past, but because it clears my head.

There's something about writing things down that slows the noise. It forces me to stop, breathe, and focus. When everything's digital—apps, alerts, tabs—it's just too easy to lose track of what really matters. Pen and paper cut through that. It becomes a quiet space to think without distraction.

I've noticed something else, too. The sharpest people I've worked with—the ones making the biggest impact—usually have some form of pen and paper at arm's length. Whether it's a legal pad, journal, or spiral notebook. Not because they can't type fast enough. It's because they know their brain works better when it's not competing with a screen.

Here's What to Automate First

Start by automating repetitive, time-consuming, and low-risk tasks. These are the items that slowly drain your day without requiring much thought: follow-up emails, appointment scheduling, data entry, file organization, and notifications.

You know, the stuff that *feels* painful. Anything that you see on your agenda and go... 'ugh...'

If you've done it more than three times, and it doesn't require serious thought, then it's a strong candidate for automation.

Next, look at areas where speed matters. If delays cost you momentum or money (things like lead response time, invoice processing, or task handoffs), automation can potentially tighten up your operational workflow. It removes the wait and keeps things moving.

Finally, ask yourself: *"What tends to break when I'm too busy to watch it closely?"* Those are your blind spots. Automating reminders, checkpoints, or status updates in these areas protects you from missed steps and overlooked details. You're not replacing yourself—you're guarding your time while ensuring accuracy and peace of mind.

What You Shouldn't Automate

Now, just because you *can* automate something doesn't mean that you *should*. Some tasks require nuanced human judgment and/or emotional intelligence. I'm talking about difficult, high-stakes conversations, strategic decisions, and anything else that might involve tone, context, or human empathy. Ultimately, if the decision changes depending on who you're talking to, it's probably not a good candidate for automation.

It sounds simple, but you also should avoid automating anything you haven't already done manually with success. If you don't fully understand the task, its logic, and edge cases, automating it simply amplifies the risk.

Finally, be careful with overautomation. When your entire system depends on dozens of interconnected tools and co-dependent triggers, one small, innocent change can create a ripple effect that breaks everything. Keep it simple. Automate what's predictable. Keep tight control over what's sensitive (or legally protected). The goal (at least initially) is to step back just far enough to focus on what only you can do.

Choose the Right Automation Tools

In tech, it's all too easy to get caught up in the latest apps, buzzwords, 'technobabble,' and popular tools. However, it's crucial to begin by identifying the specific task you want to eliminate. Every good tool solves a well-defined problem. If you can't explain the problem clearly, you won't pick the right solution—you'll just add complexity and frustration.

> *"No problem can be solved from the same level of consciousness that created it."*
>
> Albert Einstein

Look for tools that are simple to use and integrate well with what you've already invested in, and can grow with you. Avoid those that require weeks of setup or attempt to do everything. You don't need an all-in-one platform if you only need one job done reliably. One-button simplicity often beats 100 options buried in menus.

Also, don't underestimate trust. Choose tools that are well-supported, actively maintained, well-respected in the industry, and have clear, simple documentation. A dead app, broken integration, or missing support team can break your entire system. And, unfortunately, bring your entire business to a screeching halt. Remember, automation should give you leverage, not create headaches.

Automation Without Oversight Is a Trap

Just because something is automated doesn't mean it's working well. *"Set-it-and-forget-it"* is a dangerous myth.

Tools break. APIs change. Logic fails silently. If you're not checking in, you're not in control—you're just hoping everything runs like it should.

You need checkpoints. Alerts, summaries, or dashboards that tell you exactly what's happening without requiring you to micromanage the entire process. When something slips, you want to know right away—not find out when a client's been waiting three days for something that was never sent.

The best automation is accountable, not invisible. You still need to lead the process. You'll still spot patterns, redirect where needed, and improve what's working. Automation without oversight doesn't give you peace of mind. It only creates blind spots.

Reflection: Why I Don't Trust "Hands-Off" Solutions

I've watched teams fall apart and million-dollar projects get cancelled because the team trusted a system more than they trusted their own instincts (and eyes). I recall a project where automated reports suddenly stopped sending. Nobody noticed for days. By the time we caught it, we were behind schedule, over budget, and answering angry emails.

It wasn't the tool's fault. It was ours. We assumed automation meant the job was handled. We stopped checking. We surrendered leadership in exchange for convenience—and we paid dearly for it.

Now, every piece of automation I run comes with a manual check-in or a flag that says, *"Here's what changed this week."* I still love what automation can do. I just don't confuse it with leadership.

Recap: Automate Intentionally

The main focus of automation isn't to do less. Instead, it's to allow you to think clearly. When used intentionally, it removes clutter, protects your time, and amplifies your ability to focus on high-value work. When used carelessly, it creates blind spots, breaks processes, and, in the end, erodes trust.

Automation is no longer optional—it's the starting point for staying competitive. You now know what to automate first, what not to automate at all, and how to choose tools that solve problems without adding complexity. Oversight still matters, even when the system is doing the heavy lifting.

Automation doesn't eliminate leadership. It frees you up to lead with sharper focus, better energy, and a clearer mind.

That's how you scale without burning out.

Chapter 7

Dominate — Multiply Your Impact Through Leverage

"Give me a lever long enough and a fulcrum on which to place it, and I shall move the world."
Archimedes

LET'S BE CLEAR. GONE are the days when you could dominate in your field by putting in more hours. Today, you dominate by extending your influence beyond the work you can do by yourself. That's called leverage. It's how top performers multiply impact, expand their reach, and build momentum that continues with or without them.

This chapter is about building leverage that scales. Not by piling on more responsibility, but by learning how to apply effort where it matters. You'll see how to leverage tools, systems, content, money, relationships, and reputation to achieve more than your time would allow alone.

Domination doesn't come from being the loudest or busiest person in the room. It comes from understanding how to make results happen without being the one completing every single task.

What Leverage Really Means (and Why It's So Misunderstood)

Most people think leverage means power, control, or some type of financial wizardry.

It's none of those.

Leverage is simply this: getting amplified results from the same or smaller amount of effort.

Sending a well-crafted email that reaches thousands is leverage in its purest form. Building a system that runs a thousand times without you is also a form of leverage. When your reputation opens doors that you didn't knock on . . . you guessed it . . . leverage.

The problem is, most people try to find leverage without fully understanding it. They buy tools they don't need. They hire help they can't adequately manage. They build platforms that they never maintain. True leverage isn't just scaling. It's scaling the right thing, in the right direction, with the right structure behind it.

The goal here isn't to simply do *more*. You're actually trying to do *less* with a greater return. That subtle shift in thinking makes a real difference.

Tool Leverage: Extending Reach Without Additional Effort

Using the right tool extends your hands. Using the wrong one keeps them tied. Tool leverage isn't having the newest software version or the biggest software tool stack. Tool leverage simply focuses on selecting tools that replace hours of effort with moments of input.

An automation platform that handles 10 tasks while you sleep is leverage. A scheduling tool that eliminates 20 emails a week is leverage. A templated workflow that reduces mistakes and speeds up delivery—that's leverage, too.

But it's important to remember that tools aren't a magic wand. They need structure. They need context. In other words, you must understand your process before delegating it to another person, process, or tool. Otherwise, the tool just becomes another source of confusion hidden behind a layer of abstraction.

The best tools disappear into the background. They support your rhythm, amplify your output, and free up your mind to focus on what you do best. That's real leverage—not working *harder*, just working *smarter*.

People Leverage: Multiply Impact Without Micromanaging

You can't possibly scale by doing everything yourself. You must give others the autonomy to act independently. That's people leverage. Trusting someone else to handle certain aspects of the process so you can stay focused on higher-value activities.

But, to be clear, this isn't delegation in the traditional sense. Most people *delegate tasks*. The winners in the age of automation *delegate outcomes*. They say, *"Here's the result I want,"* not *"Here's how I want every step performed."* That frees your team to think, solve, and execute without needing your constant oversight.

Micromanagement kills leverage.

Okay. Let me repeat that.

Micromanagement kills leverage.

It turns talented, creative people into mere 'button-pushers.' If you hire help but still act like you're the only capable brain in the room, you've gained nothing except payroll.

> *"Micromanagement is the art of extinguishing the spark of passion, leaving only the ashes of conformity and compliance"*
>
> Anjan Pathak

People only become leverage when you teach, train, trust, and then step out of the way. When they can take an idea and run with it, you've created freedom—not just for them, but for you too.

System Leverage: Build It Once, Use It Forever

A system is a process that runs without needing to be re-engineered every time it is used. It's leverage in its purest form: do the thinking once, capture it, and let it run again and again with minimal effort.

Every mundane, repeatable task you do is an opportunity to create a system. Whether it's how you onboard clients, publish content, handle requests, or manage follow-ups—if it works, it should be built into something that doesn't rely on memory or mood.

Systems don't have to be complex. In fact, the simpler, the better. A checklist that prevents mistakes is a system. A reusable template is a system. An FAQ that stops you from answering the same question for the tenth time is a system.

You stop being the bottleneck when you start thinking in systems. That's also when knowledge evolves into an asset.

Content Leverage: Teach Once, Influence Many

One conversation helps one person. One piece of content can help thousands of people. That's the difference between direct effort and leveraged effort. Content is one of the most powerful forms of leverage you have.

Write it once. Say it once. Record it once. If it answers a question, teaches a concept, or shares a strategy, it becomes an asset that can be reused, repurposed, and resold. Every time someone reads it, watches it, or shares it, your influence spreads—without you lifting a finger.

That's how winners extend their reputation, authority, and value. Not by being everywhere in person, but by putting their thinking into formats that scale: guides, videos, articles, frameworks, and templates.

If your brain holds some knowledge that others need, capture it. Publish it. Let it work for you. The internet never sleeps, and good content has no expiration date.

Money Leverage: Buy Time, Don't Just Spend It

Most people use money to buy *stuff*. High performers use money to buy back *time*.

Money leverage means spending intentionally to free yourself from low-value tasks, unnecessary stress, or time drains that pull your focus away from what actually moves the needle. This might mean hiring someone to handle admin work, paying for faster tools, or outsourcing something that takes you 5 hours and someone else 30 minutes.

The mindset shift here is simple: stop asking, *"What does this cost?"* and start asking, *"What does this save or produce?"* Money that buys clarity, speed, focus, or more leverage is always *money* well spent.

To truly dominate, you can't cling to every dollar while wasting hours on tasks beneath your talent level. Leverage is about creating space, and money is one of the fastest, most efficient ways to do it.

Reflection: Why I Don't Shop for Groceries Anymore

When I was younger, I didn't think twice about spending an hour or more grocery shopping. Walking up and down the aisles, waiting in line, and driving home with bags of food felt normal. However, I now see things much differently.

Truthfully, I haven't physically shopped for groceries in years. Not because I can't, but because I won't spend two hours on a task that can be solved in 10 clicks. That time is better spent thinking, planning, or creating something with real return (like, say, writing this book).

Now, I'm not rich or bougie. I'm also not above running errands. I simply understand my value. Winners guard their time because they know it's the one resource they can't get back. Saving $12 on groceries doesn't make up for the $1,200 idea you didn't have because you were in line at checkout.

Time isn't just money. It's opportunity. And I don't trade mine lightly.

Reputation Leverage: When Your Name Works Harder Than You Do

Your reputation is the invisible engine behind new opportunities. When people place their trust behind your name, brand, or logo, you don't have to chase every deal, explain every decision, or prove your worth from scratch.

Your reputation does it automatically.

Of course, this type of leverage isn't built overnight. It's built over time with consistency, clarity, and character. When people know exactly what you stand for, what you deliver, and how you work, they remember you.

They recommend you.

They refer you.

> *"Your brand is what people say about you when you're not in the room."*
>
> Jeff Bezos

Now, I'm not talking about bragging or shameless self-promotion. I'm talking about alignment. Alignment between your words and your execution.

For instance, people ask: *"Do your actions match your message?" "Do your results match your promises?"* When you (and, more importantly, your customers) consistently answer *"yes"* to these questions, the more leverage your name holds in the marketplace. When your name and reputation earn people's trust before you even speak, that's domination.

Platform Leverage: Building for Payback

Alright. Let's be honest. We generally don't think of a platform as a device for creating leverage. But if your reputation is your name, then your platform is your voice. It's how you stay visible without being present. This is how people find you, learn from you, trust you, and buy from you—without even meeting you.

Whether it's your website, YouTube channel, newsletter, podcast, or community, your platform is your home base. It holds your ideas, offers, message, and social proof. And once it's built, it continues to work even while you sleep.

Platform leverage doesn't happen by accident. You build it brick by brick, piece by piece. You show up. You share value. You create things that solve problems and answer questions. Over time, these efforts compound.

A strong platform turns complete strangers into buyers, clients, partners, and brand advocates. Don't get caught up in the minutiae, such as follower count and views. This type of influence doesn't need your constant presence to grow.

Self-Leverage: Habits That Create Separation

Before you can effectively leverage tools, people, or platforms, you have to leverage yourself. That means building the internal routines that separate top performers from everyone else: clarity, discipline, decisiveness, and follow-through.

Self-leverage is waking up with a plan instead of reacting all day. It's saying "no" to what doesn't matter, so you can double down on what does. It's choosing deep work over distraction, direction over noise, and progress over perfection.

Winners don't rely on 'motivation.' They rely on systems. They train their minds to focus, their bodies to hold the pace, and their calendars to reflect real priorities. That creates separation—not just from others, but from the past version of yourself.

If you can't lead yourself, you can't lead anything.

Leverage starts from within.

Recap: Dominate Doesn't Mean Do It Yourself

To dominate in the age of AI, you don't need to do more. You need to multiply your impact with smarter leverage. That starts with understanding the difference between working harder and working higher.

We've discussed using tools to offload manual work, how to build systems that run autonomously, and how to trust people with outcomes—not just tasks. You've seen how content can scale your thinking, how money can buy back time, and how platform and reputation become assets that work long after you've stopped grinding.

But the real engine behind all of this is self-leverage. Your habits, your discipline, your clarity. Without these in place, nothing scales. With it, everything does.

Domination isn't about being the busiest. It's about being the most focused, the most intentional, and the most strategic with where and how you apply effort.

That's how you move from doing work to directing outcomes.

From being busy to becoming powerful.

Chapter 8
The Final Shift From Worker to Architect

"The best way to predict the future is to create it."

Peter Drucker

THIS IS THE POINT where we separate those who merely survive in the age of AI from those who shape it.

Up to now, we've broken down why mindless execution is merely a race to the bottom. You've seen why productivity hacks, flashy tools, and endless task lists just won't cut it. We've discussed how systems, orchestration, and strategic thinking are the real currency of modern work.

But now, it's time to go one step further.

In this chapter, you'll fully step into the role of an architect. That is, someone who doesn't just respond to the world around them. They redesign it.

Even in the age of AI, workers are necessary. Every orchestra needs instruments. But the conductor decides which instruments get played and when. Likewise, the architect decides what gets built. That's your role now. Not to keep on grinding inside the gears of someone else's machine, but to innovate. Design something smarter, but doing so on your terms. With intention. And with a vision others can't see (yet).

Workers Wait to Be Told. Architects Say What Gets Done.

There's a simple difference between workers and architects. One follows instructions. The other creates them.

Most workers spend most of their day asking: *"What should I do next?"* They wait for direction. They respond to what's already been put in motion. Even though they're talented, they're still reacting to the priorities that someone else sets for them.

Architects operate from a different place. They start by asking: *"What's missing?"* or *"What should exist today that doesn't?"* They don't wait for a plan; they create one. They don't execute tasks. They decide what the tasks are, and when they should occur. That mental leap: owning the outcome instead of just completing the steps makes all the difference.

The AI era won't be kind to passive workers. The more reactive you are, the more easily you'll be replaced. Execution is easy to automate. What's harder to replace is the person who decides what matters in the first place.

That's the architect.

The designer.

The person who can think several moves ahead.

You don't need to be a CEO or top-level executive to think this way. You don't need a title, a team, or a permission slip. What you need is ownership. You need the courage to stop waiting for someone to hand you a blueprint—and start drawing your own.

Architects don't wait to be told what matters. They decide what's missing—and start building it. That's where your value comes from now.

You Don't Need Permission to Think Big

Most people stay small because they think someone has to give them the "green light" to think big.

They wait silently for a promotion or fancier job title. They delay their dreams in the hopes that someone "higher up the food chain" will provide validation for their ideas. They wait for someone in authority to say, *"You're allowed to think like a strategist now."* But here's the truth—no one's coming. No one's going to wave a magic wand and grant you permission to level up your thinking. That's a gift you have to give yourself.

> *"Often in the real world, it's not the smart who get ahead, but the bold."*
>
> Robert T. Kiyosaki

Thinking like an architect starts with stepping outside the limits of your current role, your current habits, and your current assumptions. So stop asking: *"What should I do today?"* And start asking: *"What could I design that would make this entire process better?"* The moment you do that, you've left the assembly line behind. You're not a task-runner anymore. You're a builder.

The systems you wish existed? Build them. The inefficiencies you see? Fix them. The future you'd rather work in? Start creating it now. It doesn't matter if you're top-level, mid-level, entry-level, self-employed, or unemployed. This shift is internal, not external. It's a mindset thing. It begins with how you perceive your role, combined with your willingness to stop waiting for instructions and start shaping outcomes.

The most valuable people in any organization, at any level, are the ones who create clarity from confusion, order where there was once chaos, and direction where there was once disharmony. None of them waited for permission to start.

Architects Aren't Just Problem Solvers, They're Problem Eliminators

Workers are generally praised for solving problems (especially the ones they didn't create). Architects are respected for ensuring that those problems don't recur. There's a fundamental difference between being someone who puts out fires and being the one who redesigns the wiring so the building never catches fire again.

When you have a worker's mindset, your thinking is stuck in triage mode. You're constantly wondering: *"What's broken?" "What's urgent?" "What problem needs my attention right now?"* That approach can be useful in short bursts, but it can't scale. It drains you because you end up spending all your time reacting, not directing.

Architects step back. They see recurring problems as signs of a deeper flaw in the system. Instead of asking, "*How do I fix this again?*" they ask, "*Why does this keep breaking?*" They design solutions that permanently remove friction. They build systems, create safeguards, and put alerts in place. They understand that one well-designed solution eliminates reactivity.

That is how you shift from effort to impact, not by doing more work, but by eliminating the need to do work in the first place.

Design Your World Before You Enter It

Most people step into a broken environment and then try to outshine everyone else by working harder. They push through inefficiencies, tolerate poorly planned systems, and adapt to chaos like it's some kind of 'badge of honor.' That's the worker mindset. They adjust themselves just enough to fit comfortably into the dysfunction.

Architects do the opposite. They don't conform to disorder—they redesign the system.

Whether it's your calendar, your workspace, your communication flow, or your client onboarding process—everything works better when it's intentionally designed. Not cluttered. Not patched together. Not just *"how it's always been done."*

Think about how much time you waste dealing with disorganized files, unclear expectations, or redundant back-and-forth banter. Now imagine if those issues were reduced, or, better yet, eliminated before you ever sat down to work. That's the difference between running in circles and moving forward with purpose.

This is the architect's advantage. Not working inside a broken system—shaping it into something better. You decide what kind of environment supports your optimal performance, then you start creating it.

Always Build With Leverage in Mind

The power to amplify results from a single action, decision, or system is leverage. It's how you escape the grind of linear effort—where one hour equals one outcome—and step into exponential impact.

In the last chapter, we touched on how leverage comes in many forms: automation, delegation, templates, workflows, compounding knowledge, and even relationships. But, the common thread is this: everything you create should free up time, amplify output, or enhance clarity. If it doesn't, it's just another task masquerading as something important.

As an architect, you have to start treating leverage as a filter. Before you take something on, ask: *"Will this scale?" "Will this make future decisions easier?" "Will this eliminate a recurring problem?" "Will this free up my energy to operate at a higher level?"*

If the answer is no, don't chase it. Replace it. Because in a world where AI, systems, and strategy can do the heavy lifting, your real value comes from thinking in big moves, not small, repetitive motions.

The Real Power Is in Direction, Not Execution

Execution matters, but it's not where true power lives. Real power lives in direction.

The value of a worker is measured by how fast they move. How quickly can they complete the task at hand? But an architect's value is intrinsically tied to the value of the final result and how they obtain it. You can be busy all month executing a poor strategy. You can sprint toward the wrong goal and still lose. Why? Because speed doesn't matter if you're headed in the wrong direction.

Conductors and architects bring something far more valuable than effort. They bring clarity. They cut through noise, align priorities, and define what matters most. They make the decisions that shape execution—not just participate in it. Their value lies in what they decide to do, when to do it, and what gets ignored entirely.

In a world filled with distractions, unclear objectives, and people chasing metrics that don't really matter, direction is everything. A person who cuts through the noise to bring clear direction becomes the anchor that others rely on. That's the difference between being a cog in the machine and the one who builds it.

Execution gets things done. Direction decides what gets done—and why.

Architecting Your Future

You need to think past the work. Focus on how you move through the world.

Being the architect of your future means refusing to let circumstance, convenience, or outdated expectations define your direction. It means recognizing that everything around you—your habits, tools, routines, priorities—can either be chosen or inherited. Architects choose. They don't coast into the future. They create it.

This also means that your career isn't a ladder anymore. It's a platform. One that you can build, expand, and reconfigure as needed. AI is your power tool. Automation, workflows, templates, outsourced labor, smart delegation aren't just efficiencies. They're leverage points. And when you stack them along with clear thinking and bold direction, you're no longer competing. You're creating.

This isn't about some 5-year plan. It's about uncompromising ownership. It's understanding that the smallest decisions today—the way you manage time, the tools you use, the mental boundaries you set—are the scaffolding for what your future will look like tomorrow.

The question isn't whether change is coming. It's whether you're going to be ready for it... and how you'll be shaped by it.

Stop asking for permission.

Stop bracing for impact.

Start laying the foundation.

Recap: Stop Working. Start Building

Congratulations. You've made the final shift.

You're not just here to finish tasks. You're here to shape outcomes. You're not waiting to be told what matters—you're deciding what gets built, how it works, and who benefits from it.

You're not playing small.

You're playing to win.

Workers focus on the next instruction. Architects focus on what's missing, what's broken, and what could exist if someone just stepped up and designed it.

The future belongs to people who can see clearly, think structurally, and build with intention. You don't need to outwork everyone. You just need to out-design the problem.

Don't ask, *"What's the next step?"*

Ask, *"What system would make this problem irrelevant?"*

Ask, *"What process would make this result automatic?"*

Ask, *"What do I want to build that no one else sees yet?"*

That's how you ***orchestrate***.

That's how you ***automate***.

That's how you ***dominate***.

Did This Book Help You?

IF THIS BOOK HELPED you feel more prepared for the future, I'd be forever grateful if you take a moment to leave a review. Your feedback doesn't just help authors like me. It helps other business owners and career-focused individuals find the tools they need to adapt, stay competitive, and thrive, no matter what the global economy throws their way.

Please leave a review on Amazon or your favorite retailer and let others know why this book matters, and how it can help them too.

Thanks again for reading.

With Love and Gratitude,

Donovan Garett

Wait... There's More!

YOU'VE JUST READ THE kind of material most people never even think about—let alone apply. That's what separates readers from winners.

If you're the kind of small business owner who builds, leads, and thinks ahead, then I invite you to join my list. You'll get early access to new releases, exclusive strategy guides, and tools to help you stay ahead of the curve—without drowning in noise, hype or fluff.

Scan the QR code to sign up in 10 seconds or less.

Made in the USA
Monee, IL
08 November 2026

67375844R10069